*Polygonum arifolium*

1/12'

*Carex albolutescens*

⅔

*Brasenia schreberi*

⅓

*Alisma subcordatum*

# A Floristic Survey

## of the Aquatic Habitats in

## Western Tazewell County,

## Virginia

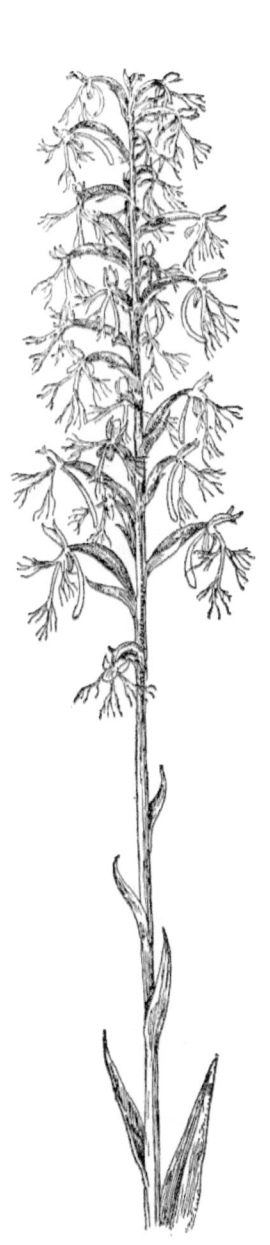

Randy F. McNew Crouse

Frontispiece: *Habenaria lacera*, Ragged Fringed Orchid.
( Illustrations from Britton & Brown and FCIT )

Cover illustrations created with images from Britton & Brown, USDA
NRCS, and FCIT

# A Floristic Survey of the Aquatic Habitats in Western Tazewell County, Virginia

~

Randy F. McNew Crouse

*Houstonia caerulea*, Bluets
Illus. courtesy FCIT

Copyright © 2012
Randy F. McNew Crouse
1116 Summerwood Cir SE
Huntsville, Alabama 35803

**First Edition, 2nd issue**

ISBN  978-1-300-15814-1

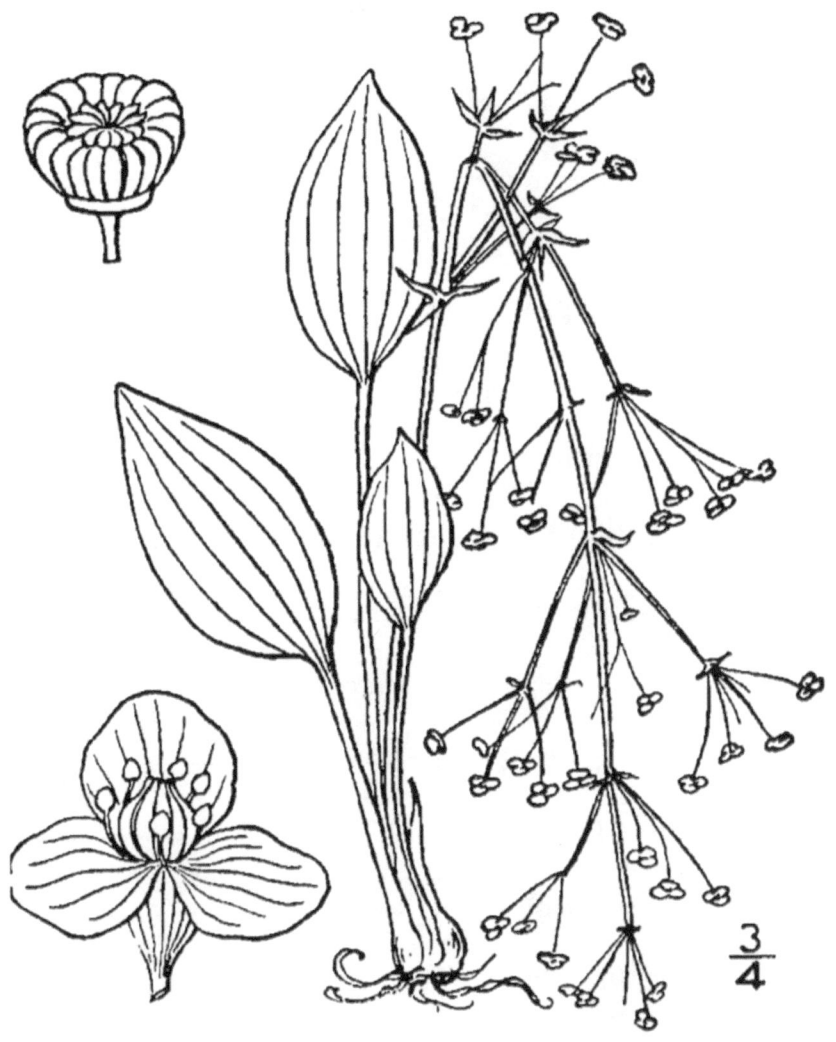

**_Alisma subcordatum Rafinesque_**
American Water Plantain
(illus. fr.Britton & Brown, 1913)

# Table of Contents

***Sagittaria latifolia* L.**
Broadleaf Arrowhead
(clipart courtesy FCIT)

# Table of Illustrations

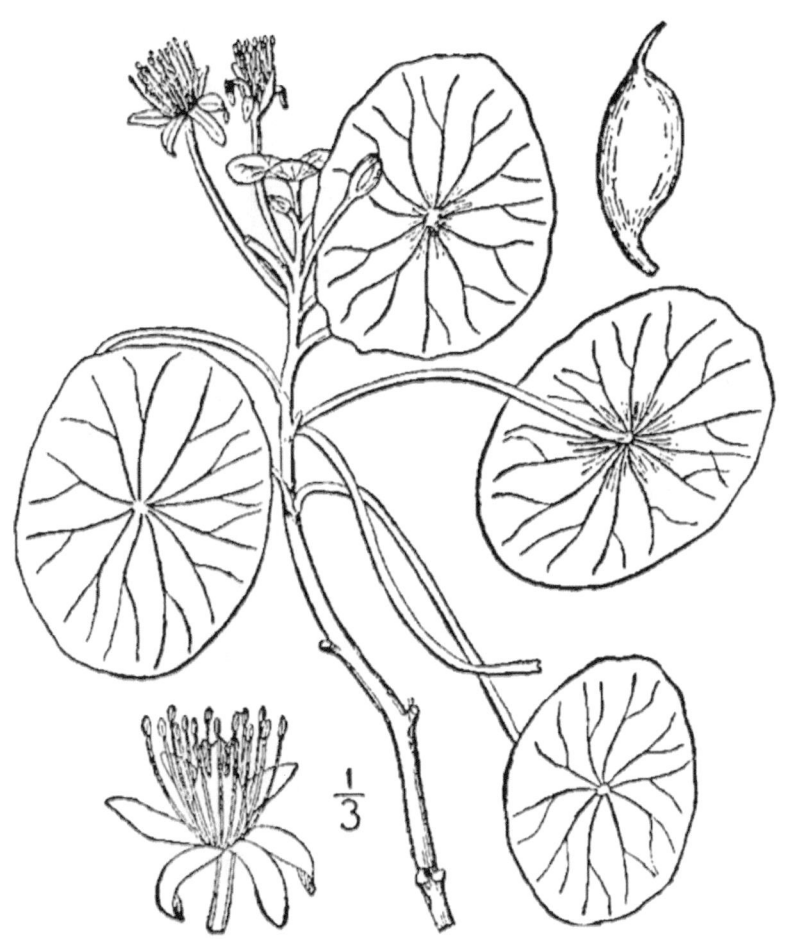

$\frac{1}{3}$

***Brasenia schreberi* J.F. Gmel**.

Watershield; Dollar Bonnett

(illus. fr.Britton & Brown, 1913)

# A Floristic Survey of the Aquatic Habitats in Western Tazewell County, Virginia.

Randy F. McNew Crouse

## ABSTRACT

The drainage patterns, topography, and climate of Tazewell County, Virginia are discussed. Aquatic plant growth factors, their importance to animals and man, and some difficulties involved in delineating the group are discussed. Aquatic habitats in the western part of Tazewell County and their associated plants are listed and unusual species or county records are cited.

## I. INTRODUCTION

Because hydrophytes indicate the condition of the water supply, they are of great environmental significance. It is hoped that this work may be useful in this respect, and in providing additional distribution data for Virginia's flora.

After reviewing the U.S. Geological Survey topographic maps, (Amonate, Richlands, Pounding Mill, Jewell Ridge, and Tazewell quadrangles and the U.S. D. A. Soil Survey map of Tazewell County, I have concluded that extensive marshes, lakes, or swamps do not exist therein. Many miles of travel have also confirmed this.

## II.    PHYSIOGRAPHY

This survey has been limited to the western half of the county; that is, west of Tazewell (see Fig. 1). For this reason, the physiographical description will be thus limited; especially in the consideration of the drainage patterns. Tazewell County occupies part of the Ridge and Valley Province and the Appalachian Plateau. On the southeast are the Appalachian Mountains, on the northeast lies the Cumberland Plateau. It is on the divide that separates the drainage of the New River and the Tennessee River. The terrain is characterized by abrupt, elongated ridges and deep narrow valleys. These ridges and valleys run in a northwest to southwest direction in conformance with the main valley.

These numerous parallel ridges contrast with the Cumberland Plateau of the northwest. This gives the region an overall uneven relief with few areas of smooth terrain. Some of the broader stretches of smooth terrain, varying from a quarter to a half mile are along Clinch River in the northwestern area, near Richlands and Doran. Consequently, this area of the county contains the largest concentration of poorly drained bottomland and hydrophytic flora.

The mountain ridges vary in elevation from 760 to 1370 meters, with some peaks higher. The valleys floors have an average elevation of 600 meters. The Tennessee River basin encompasses the central, western, and southern area. The Clinch River and its tributaries drain about 275 square miles of the county. Dry Fork basin, on the north, flows northward into West Virginia and finally to the Ohio River.

## III. CLIMATE

The climate of Tazewell County is not modified by oceanic influences. A great deal of variation occurs in seasonal temperatures, and there is a notable difference in the summer maximum and the winter minimum. The average length of the growing season is 169 days. U.S. Weather Bureau data taken at Bluefield, West Virginia, which borders Tazewell County on the northeast, is given in Table I., below. (Porter, 1948 and Wagner, 2012)

TABLE I. U.S.W.B. Data

| Temperature (°F) | | | Precipitation (inches) | | | |
|---|---|---|---|---|---|---|
| Mean | Max | Min | Mean | Driest Year | Wettest Year | Avg. Snowfall |
| 53.8 | 98 | -26 | 37.22 | 27.00 | 60.03 | 36.1 |

The climate of Tazewell County is described by R. C. Jurrey, Bureau of Plant Industry, Soils, and Agriculture, as being

"healthful and favorable to the local types of agriculture".
(Corn, wheat, oats, and hay are the leading crops.) (Porter,
1948) It seems therefore, that this climate would be healthful
and favorable to a great variety of wild vegetation. Although
extensive lakes do not exist here; wherever water may be
found in plentiful supply year-round, hydrophytes will be
also.

## IV.    AQUATIC PLANTS

Hydrophytes are given many definitions. "Plants that grow
with a part or all of the body in mud or water." (Prescott,
1980) "Plants which a typical terrestrial taxonomist usually
does not collect for fear of getting his feet wet." (Beal, 1977)
These are fair definitions but there are in, my opinion, no
group of plant species which can be called a complete list of
aquatic flora. Vascular plants exist in nearly every
conceivable environment and therefore in an infinite variety
of moisture conditions. This creates a continuum of plant
species which can increasingly tolerate less water. The
question is, where to draw the line. This is obviously every

author's responsibility and each chooses a different "line". I have deferred to the expertise of Robert K. Godfrey and Jean W. Wooten (1981) in the determination of aquatic species. At any rate, the selection of an aquatic flora is completely arbitrary.

Several factors have influenced the structure of aquatic plants, and the species able to adjust have done well. As much as 75% of the light that reaches the surface of water is reflected. Some submersed plants have numerous and small finely dissected leaves to collect more light. Another development is seen in the (Potamogetonaceae) elongation of the petioles and stem internodes. (Prescott, 1980)

Gas exchange is accomplished through the development of arenchyma providing an extensive cell surface/internal atmosphere interface. (Beal, 1977) Nutrients and essential elements pose no problem as they are in solution, and frequently more than the root system is bathed in it. Consequently, many aquatic plants have weakly developed xylem and phloem. In addition, the sclerenchyma and collenchyma are often poorly developed where support may be a hinderance.

Reproduction may also be & problem as it is entirely dependent upon water currents. By and large, most of the plants are able to reproduce asexually, and rhizomes proliferate readily. (Prescott, 1980)

Although the above characteristics pertain to those plants on the more aquatic end of the spectrum; they do in some degree affect semi-aquatics as well. The major influence, however, affecting the distribution of aquatic plants is obviously the distribution of water. This makes the study of hydrophytes a unique and important undertaking. Aquatic plants may seem insignificant and useless to some, especially boaters and fishermen. They do, nevertheless, hold an important niche in the ecosystem. Aquatic, semi-terrestrial, and terrestrial animals such as the muskrat and moose depend largely on aquatic plants as a food source. Many species of birds directly or indirectly rely on aquatic plants. Fish either use them directly or sustain themselves upon lesser creatures that live amongst them. Aquatic plants play a limnological role in beach building, the filling in of lake margins with

eutrophication and thereby preventing shore erosion.
(Prescott,1980)

Aquatic plants are a source of evolutionary data. "Most plants in the water have about the same relationship to terrestrial relatives that aquatic mammals such as whales and dolphins have to land animals."(Prescott, 1980)  That is, they have developed on land and have "returned" to the water. Increasing interest is being shown toward aquatic biota in regard to water quality and quantity. Because species of most plants are highly selective of their habitat, a population survey may be used as an indicator of physical-chemical conditions in their habitat.

Tazewell County, a county of mountains and valleys with many streams and forests at a variety of altitudes; offers a diverse flora to the botanist or anyone who appreciates nature. "Many have been the pleasant days which I have spent in botanical rambles on these mountains where from frost 'til frost flowers are ever found." (Bickley,1852)

***Lobelia cardinalis (L.)***
Cardinal Flower
Illus. courtesy FCIT

# V.    LIST OF AQUATIC FLORA

The following list, I submit as the Aquatic Flora of Tazewell
County, Virginia. An asterisk (*) indicates that this specimen is
a county record determined by A.M. Harvill and submitted for
addition to his Atlas of the Virginia Flora (1977, 1985). The
remainder of the specimens listed was taken from the
aforementioned work. The numbers inside the triangle symbol
refer to the collection site. The included map (Fig. 1.) and
legend (Table 2. and 3.) should be used in conjunction with the
list to obtain elevations, detailed location descriptions, soil
types, habitat data and associated plants.

Collections were made during the 1983 and 1984 growing

season from April 28 through September 7. ( Crouse, 1984 )

A total of two hundred sixty-eight species representing seventy-

three families now make up this list. It is comprised as follows:


**Monocotyledons**

18 Families

48 Genera

102 Species

**Dicotyledons**

55 Families

112 Genera

166 Species

# VI. COLLECTION SITE MAP AND TABLES

Table 2. Symbol Legend for site identification and county record indicators.

| Symbol | Meaning |
|--------|---------|
| $\triangle$ n | Numeral inside gives site number from Figure 1, data in Table 3. |
| * | Indicates a county record |
| $\triangle$ 2 A | A subscript letter refers to several sites designated by one number on Figure 1, due to their close proximity. Table 3 treats these separately. |

A 57.68: H 99/185

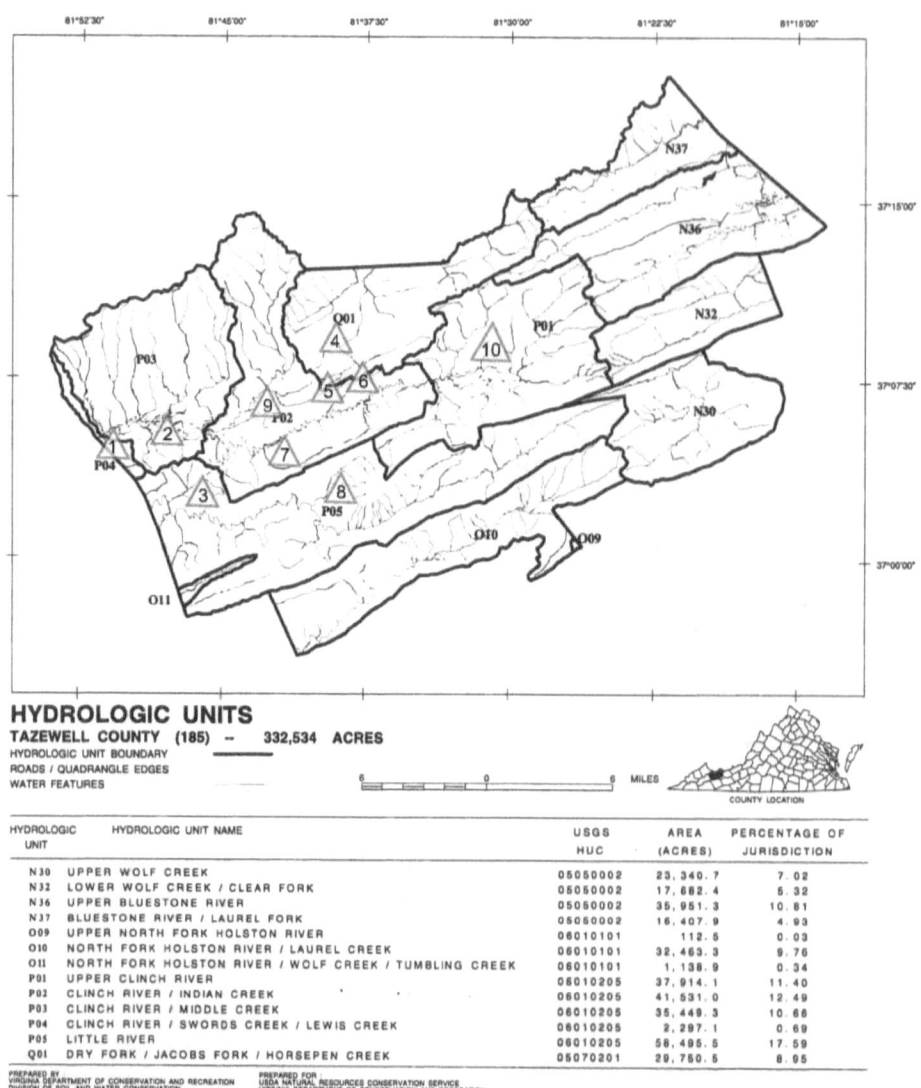

## HYDROLOGIC UNITS
**TAZEWELL COUNTY (185)  —  332,534 ACRES**
HYDROLOGIC UNIT BOUNDARY
ROADS / QUADRANGLE EDGES
WATER FEATURES

| HYDROLOGIC UNIT | HYDROLOGIC UNIT NAME | USGS HUC | AREA (ACRES) | PERCENTAGE OF JURISDICTION |
|---|---|---|---|---|
| N30 | UPPER WOLF CREEK | 05050002 | 23,340.7 | 7.02 |
| N32 | LOWER WOLF CREEK / CLEAR FORK | 05050002 | 17,882.4 | 5.32 |
| N36 | UPPER BLUESTONE RIVER | 05050002 | 35,951.3 | 10.81 |
| N37 | BLUESTONE RIVER / LAUREL FORK | 05050002 | 16,407.9 | 4.93 |
| O09 | UPPER NORTH FORK HOLSTON RIVER | 06010101 | 112.5 | 0.03 |
| O10 | NORTH FORK HOLSTON RIVER / LAUREL CREEK | 06010101 | 32,463.3 | 9.76 |
| O11 | NORTH FORK HOLSTON RIVER / WOLF CREEK / TUMBLING CREEK | 06010101 | 1,138.9 | 0.34 |
| P01 | UPPER CLINCH RIVER | 06010205 | 37,914.1 | 11.40 |
| P02 | CLINCH RIVER / INDIAN CREEK | 06010205 | 41,531.0 | 12.49 |
| P03 | CLINCH RIVER / MIDDLE CREEK | 06010205 | 35,449.3 | 10.66 |
| P04 | CLINCH RIVER / SWORDS CREEK / LEWIS CREEK | 06010205 | 2,297.1 | 0.69 |
| P05 | LITTLE RIVER | 06010205 | 58,495.5 | 17.59 |
| Q01 | DRY FORK / JACOBS FORK / HORSEPEN CREEK | 05070201 | 29,750.5 | 8.95 |

PREPARED BY
VIRGINIA DEPARTMENT OF CONSERVATION AND RECREATION
DIVISION OF SOIL AND WATER CONSERVATION
RICHMOND, VIRGINIA

PREPARED FOR
USDA NATURAL RESOURCES CONSERVATION SERVICE
VIRGINIA DEPARTMENT OF CONSERVATION AND RECREATION
VIRGINIA DEPARTMENT OF ENVIRONMENTAL QUALITY

USDA NRCS NATIONAL CARTOGRAPHIC AND GIS CENTER, FT. WORTH, TX, 1995

REVISED JANUARY 1995 100504

**Figure 1. Collection Site Map.** Collection sites are indicated by numbered triangles. Site 2 contains three individual sites; A, B and C. See Table 3 for site description and soil units.

## TABLE 3 COLLECTION SITE DATA (See Figure 1.)

| SITE NO. | ELEVATION (METERS) | SOIL UNIT TYPE | LOCATION | HABITAT | ASSOCIATED GENERA |
|---|---|---|---|---|---|
| 1 | 580 | 32A | Doran Bottom, Rt. 743 Prop. of James Reynolds | Wooded swamp | Eleocharis, Carex, Alisma, Ilex, Cephelanthus, Polygonum, Rubus, Nyssa, Quercus, Sphagnum. |
| 2-A | 610 | 35A, 45A | Page St. behind Jack Sykes Cadillac, Richlands | Open Marsh | Juncus, Mimulus, Cornus, Typha, Urtica, Thalictrum, Carex, Mentha, Polygonun, Scirpus, Asclepius |
| 2-B | 610 | 50 | Int. of Farmer St. and Rt. 609 (Kent's Ridge Road), Richlands. | Open marsh, now partially drained or overfilled | Myosotis, Typha, Humulus, Acorus, Scirpus, Sagittaria, Habenaria, Verbena, Polygala, Cornus, Rosa. |
| 2-C | 610 | 50 | Vacant lot, 2411 S. First St., Richlands. | Spring, in field | Cuscuta, Thalictrum, Typha, Rosa, Carex, Sambucus, Trifolium, Acorus |
| 3 | 700 | 32A | Brown Hollow at Rt. 739 off of Rt 19; between Claypool Hill & Wardell. | Large marsh, app. 10 acres | Scirpus, Helenium, Acorus, Typha, Polygonum, Juncus, Sagittaria, Lobelia, Myosotis, Verbena. |
| 4 | 550 | 18B | Int. of Rt. 637 & Rt. 626. Dry Fork. | Spring and stream | Veronica, Carex, Equisetum |
| 5 | 730 | 32A, 35A | Rt. 686, McGuire Valley | Stream and marsh | Scirpus, Polygonum, Juncus, Carex, Hypericum, Veronica, Eleocharis |
| 6 | 670 | 12E, 20C, 21E | Rt. 639; Cliffield | Small marsh | Typha, Juncus, Hypericum |
| 7 | 610-950 | 12E, 53F, 38E, 10E, 51E | Paint Lick Mountain | Spring, field, and woods | Carex, Saxifriga, Asparagus, Mentha, Scirpus, Cyperus. |
| 8 | 760-1350 | 38E, 53F, 10E, 51E, 40D | Morris Knob | Open field, woods, heath bald | Maianthemum, Houstonia, Carex, Trillium, Urtica, Impatiens ,Kalmia, Rhododendron, Saxifriga, Viburnum, Triosteum, Malva. |
| 9 | 660 | 12E, 20C, 20D | Rt. 631, 3.2 kM. S.W. of Busthead Mt.; C. Bandy property | Farm pond and marsh | Potamogeton, Lysimachia, Sium, Onoclea, Sabatia, Agrimonia, Carex, Juncus, Myosotis, Mentha, Cornus, Sisyrinchium |
| 10 | 780 | 5E, 53F | Lincolnshire Park Lake, Tazewell, Va. | Edge of Small Lake | Typha, Juncus, Scirpus, Carex, Ranunculus |

**Soil Unit Legend**
5E Bland-Rock outcrop complex, 25 to 50 percent slopes
10E Calvin channery silt loam, 35 to 55 percent slopes, very stony
12E Carbo-Rock outcrop complex, 25 to 65 percent slopes
20C Frederick silt loam, 7 to 15 percent slopes
20D Frederick silt loam, 15 to 25 percent slopes
21E Frederick gravelly silt loam, 25 to 35 percent slopes
18B Craigsville very gravelly sandy loam, 0 to 5 percent slopes, frequently flooded

**Soil Unit Legend**
32A Melvin silt loam, 0 to 2 percent slopes, frequently flooded
35A Newark-Lindside complex, 0 to 3 percent slopes, occasionally flooded
38E Oriskany gravelly fine sandy loam, 35 to 55 percent slopes, extremely stony
40D Paddyknob gravelly loam, 15 to 35 percent slopes, very stony
45A Pope fine sandy loam, 0 to 2 percent slopes, rarely flooded
50 Udorthents-Urban land complex
51E Wallen-Rock outcrop complex, 35 to 80 percent slopes, extremely stony
53F Westmoreland-Poplimento-Berks complex, 35 to 65 percent slope

***Osmunda cinnamomea L.***
Cinnamon Fern
(clipart courtesy FCIT)

# The Aquatic Flora of Tazewell County, Virginia

## Pteridophytes through Monocotyledons

### Equisetaceae

 Equisetum arvense L.

 * E. hyemale L.

### Lycopodiaceae

Lycopodium flabelliforme (Fernald) Blanchard

L. lucidulum Michaux

L. obscurum L.

L. obscurum var. dendroideum (Michaux) D. Eaton

### Ophioglossaceae

Botrychium virginium (L.) Swartz

### Osmundaceae

 Osmunda cinnamomea L.

 * O. regalis L.

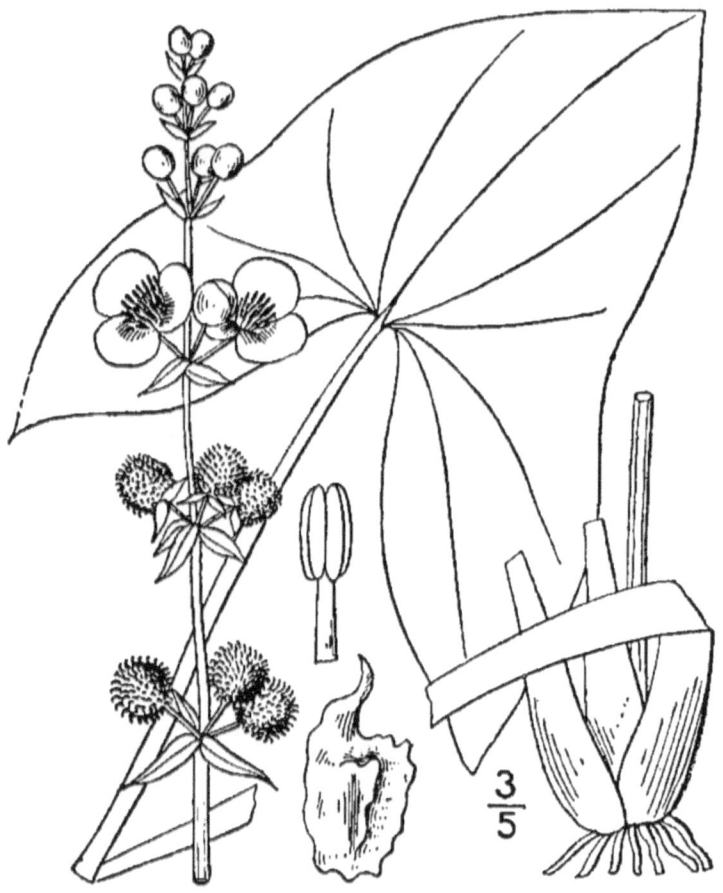

**Sagittaria longirostra (Micheli) J.G. Smith**
Duck Potato, Longbeak Arrowhead
(syn. Sagittaria australis)
(illus. fr.Britton & Brown, 1913)

**Polypodiaceae**

Adiantum pedatum L.

Asplenium trichomanes L.

 Athyrium asplenioides (Michaux) A. Eaton

A. thelypteroides (Michaux) Desv.

Dryopteris campyloptera Clarkson

D. intermedia (Willdenow) Gray

 Onoclea sensibilis L.

Thelypteris hexagonoptera (Michaux) Fee

T. noveboracensis (L.) Nieuwland

**Alismataceae**

* Alisma subcordatum Raf.

Sagittaria latifolia Willdenow

* S. longirostra (Micheli) J. G. Smith

***Carex albolutescens Schweinitz***
Greenish-white Sedge, Pale Straw Sedge
(illus. fr.Britton & Brown, 1913)

**Amaryllidaceae**

Hypoxis hirsuta (L.) Coville

**Araceae**

 B Acorus calamus L.

 Arisaema triphyllum (L.) Schott.

**Cyperaceae**

 * A Carex albolutescens Schweinitz

C. amphibola Steudel

C. baileyi Britton

C. bromoides Willdenow

C. caroliniana Schweinitz

C. complanata Torrey &Hooker

C. crinita Lam.

 * A C. cristatella Britton

C. debilis Michaux

C. frankii Knuth

 * C. granularis Muhlenburg

### *Carex seorsa Howe*
Swamp Star Sedge, Weak Stellate Sedge
(illus. Harry Charles Creutzburg fom McKenzie 1940)

C. grisea Wahlenberg

C. intumescens Rudge

C. leptalea Wahlenberg

 * C. lupulina Willdenow .

C. lurida Wahlenberg

C. normalis Mackenzie

C. prasina Wahlenberg

 * C. seorsa Howe

C. scoparia Willdenow

C. stipata Willdenow

 C. tribuloides Wahlenberg

Carex vulpinoidea Michaux

Cyperus rivularis Knuth

 C. strigosus L.

 A Eleocharis erythropoda Steudel

E. obtusa (Willdenow) Schultes

 * B E. tenuis (Willdenow) Schultes

**Sisyrinchium atlanticum Bicknell**
Narrow-leaved Blue-eyed Grass
(illus. fr. Britton & Brown, 1913)

Rhyncospora capitellata (Michaux) Vahl

 Scirpus atrovirens Willdenow

S. cyperinus (L.) Knuth

 * S. pendulus Muhlenberg

S. polyphyllus Vahl

 * S. validus Vahl

## Iridaceae

 Iris pseudoacorus L.

 Sisyrinchium angustifolium Miller

 * S. atlanticum Bicknell

## Juncaceae

Juncus acuminatus Michaux

J. diffusissimus Buckley

J. effusus L.

J. tenuis Willdenow

Luzula acuminata Raf.

 L. echinata (Small) Hermann

***Medeola virginica (L.)***
Indian Cucumber
(clipart courtesy FCIT)

**Lemnaceae**

Spirodella polyrhiza (L.) Schleiden

**Liliaceae**

Aletris farinosa L.

Lillium michauxii Poiret

Lillium superbum L.

 Medeola virginiana L.

Smilax glauca Walter

S. rotundifolia L.

**Orchidaceae  (Habenaria = Platanthera)**

Aplectrum hyemale (Willdenow) Torrey

Calopogon tuberosus (L.) Bsp.

Habenaria ciliaris (L.) R. Brown

H. flava (L.) Sprengel

 H. lacera (Michaux) Lodd.

Listera smallii Wiegand

Spiranthes cernua (L.) Richard

***Glyceria septentrionalis* Hitchcock**
Floating mannagrass
(illus. fr. Britton & Brown, 1913)

S. gracilis (Bigelow) Beck

**Poaceae**

Agrostris perennans (Walter) Tuckerman

Andropogon scoparius Michaux

 * A. virginicus L.

Cinna arundinacea L.

 Echinochloa crusgalli (L. )Beauvois

Elymus riparius Wiegand

E. villosus Willdenow

E. virginicus L.

Festuca obtusa Biehler

Glyceria melicaria (Michaua) Huhbard

 * G. septentrionalis Hitchcock

G. striata (Lam.) Hitchcock

Leersia oryzoides (L.) Swartz

Muhlenbergia frondosa (Poiret) Fernald

Panicum dichotomiflorum Michaux

Paspalum laeve Michau

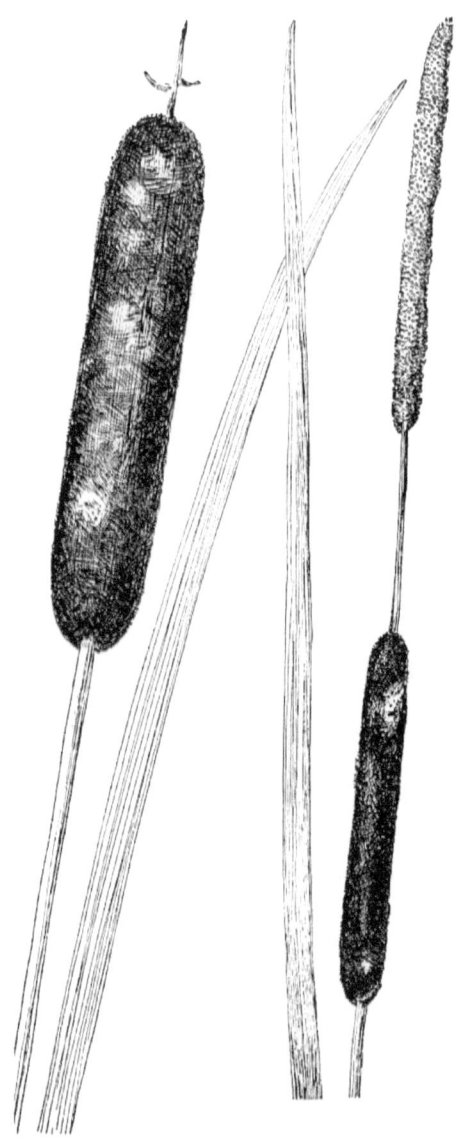

***Typha latifolia (L.)***     ***Typha angustifolia (L.)***
Cat-tail                    Narrow-leaved Cat-tail
(illus. courtesy FCIT)

## Potamogetonaceae

 * Potamogeton nodosus Poiret

## Sparganiaceae

Sparganium americanum Nuttall

## Typhaceae

 * Typha angustifolia L.
B

 T. latifolia L.

***Potamogeton nodosus Poiret***
Long-leaf Pondweed
(illus. fr. Britton & Brown, 1913)

**Sium suave Walter**
Water Parsley
(illus. fr. USDA NRCS)

# The Aquatic Flora of Tazewell County, Virginia

## Dicotyledons

**Aceraceae**

> Acer negundo L.

> A. rubrum L.

>  A. saccharinum L.

**Anacardiaceae**

>  Rhus radicans L.

**Apiaceae**

>  Cicuta maculata L.

> Conium maculatum L.

> Cryptotaeania canadensis (L.) Dc.

> Heracleum lanatum Michaux

> Zizia trifoliata (Michaux) Fernald

>  * Sium suave Walter

**Aquifoliaceae**

>  * Ilex opaca Aiton

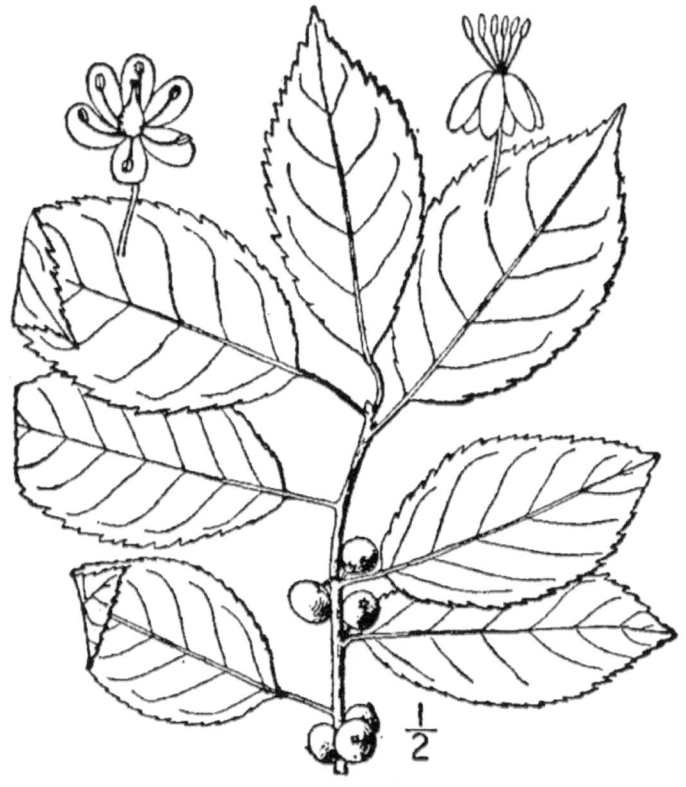

***Ilex verticillata (L.) Gray***
Swamp Holly, Winterberry, Black Alder Winterberry
(illus. fr. Britton & Brown, 1913)

\* I. verticillata (L.) Gray

## Aristolochiaceae

 Aristolochia macrophylla Lam.

## Asclepiadaceae

 Asclepias incarnata L.

## Asteraceae

Aster cordifolius L.

A. novae-angliae L.

A. puniceus L.

A. sagittifolius Willdenow

A. umbellatus Miller

**Bidens cernua (L.)**

**B. frondosa**

Conoclinium coelestinun (L.) Dc.

Erigeron philadelphicus L.

E. pulchellus Michaux

Eupatorium perfoliatum L.

E. serotinum Michaux

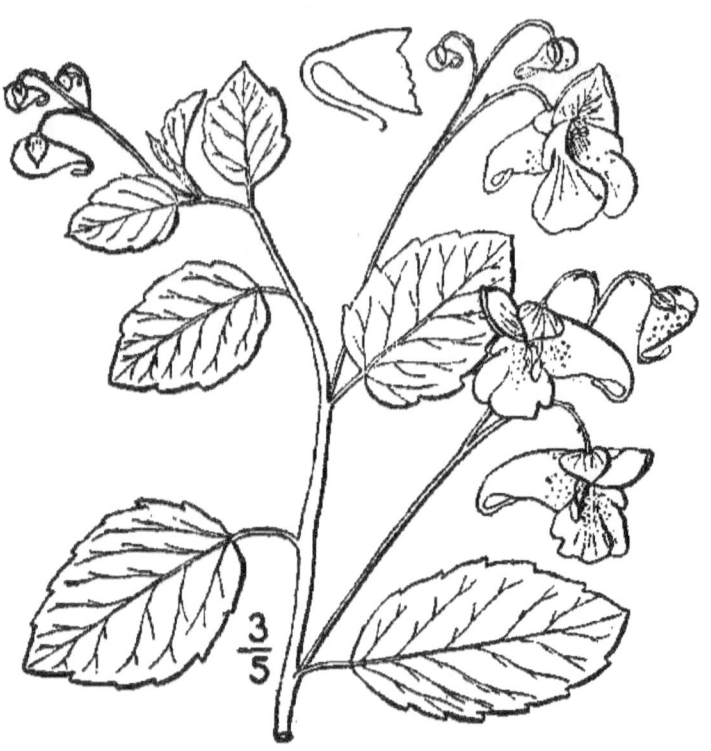

***Impatiens capensis Meerburg***
Jewel Weed, Touch-me-not
(illus. fr. Britton & Brown, 1913)

Helenium autumnale L.

Rudbeckia laciniata L.

 Senecio aureus L.

 Solidago altissima L.

S. patula Willdenow

Verbesina alternifolia (L.) Britton

Vernonia gigantea (Walter) Trelease

 V. noveboracensis (L.) Michaux

Xanthum strumarium (L.)

## Balsaminaceae

 Impatiens capensis Meerburg

 I. pallida Nuttall

## Betulaceae

Alnus serrulata (Aiton) Willdenow

 Carpinus caroliniana Walter

## Boraginaceae

 Myosotis laxa Lehmann

***Callitriche heterophylla Pursh***
Two-headed Water Starwort
(illus. fr. USDA NRCS)

## Brassicaceae

 \* Cardamine bulbosa (Schreber) Bsp.

C. pennsylvanica Willdenow

C. rotundifolia Michaux

Dentaria laciniata Willdenow

 Nasturtium officinale R. Brown

## Cabombaceae

 \* Brasenia schreberi J.F. Gmel.

## Callitrichaceae

 \* Callitriche heterophylla Pursh

## Campanulaceae

 Campanula americana L.

 Lobelia cardinalis L.

 L. inflata L.

 L. siphilitica L.

L. spicata Lam.

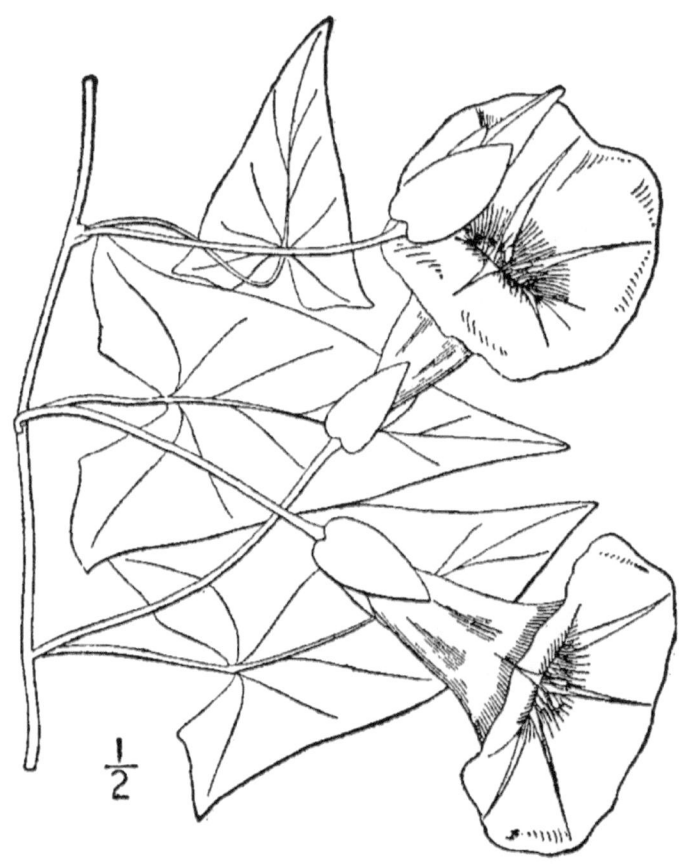

**Calystegia sepium (L.) R. Brown**
Hedge False Bindweed
(syn. Convolvulus sepium)
(illus. fr. Britton & Brown, 1913)

## Caprifoliaceae

Lonicera japonica Thunberg

 Sambucus canadensis L.

Viburnum cassinoides L.

## Cornaceae

 Cornus ammomum Miller

## Caryophyllaceae

Stellaria graminea L.

## Convulvulaceae

 * Calystegia sepium (L.) R. Brown

 * Cuscuta campestris Yuncker

Ipomoea purpurea (L.) Roth

## Crassulaceae

 Penthorum sedoides L

## Cucurbitaceae

Sicyos angulatus L.

*Chimaphila maculata (L.) Pursh*
Spotted Wintergreen
(illus. fr. Britton & Brown, 1913)

## Ericaceae

 Chimaphila maculata Pursh

Gaylussacia frondosa (L.) Torrey & Gray .

Lyonia ligustrina (L.) Dc.

Rhododendron nudiflorum (L.) Torrey

R. viscosum (L.) Torrey

Vaccinium atrococcum (Gray) Heller

Zenobia pulverulenta (Bartram) Pollard

## Euphorhiaceae

Acalypha rhomboidea Raf.

Euphorbia maculata L.

E. nutans Lagasca

## Fabaceae

Apios americana Medicus

## Fagaceae

Quercus prinus L.

***Mentha spicata L.***
Spearmint

***Mentha piperita L.***
Peppermint

(illus. courtesy FCIT)

## Gentianaceae

Bartonia virginica (L.) Bsp.

 Sabatia angularis (L.) Pursh

## Hamamelidaceae

 Hamamelis virginiana L.

## Hypericeae

 Hypericum mutilum L.

## Lamiaceae

Dracocephalum virginianum L.

 Lycopus americanus Barton

 * L. virginicus L.

 Mentha piperita L.

 * M. spicata L.

Pycnanthemum tenuifolium (L.) Michaux

Scuttelaria laterifolia L.

Stachys tenuifolia Willdenow

 Teucrium canadense L.

***Cuphea viscosissima Jacquin***
Blue Waxweed
(clipart courtesy FCIT)

**Lauraceae**

Lindera benzoin (L.) Blume

**Linaceae**

Linum striatum Walter

L. virginianum L.

**Lythraceae**

 Cuphea viscosissima Jacquin

**Magnoliaceae**

 Liriodendron tulipifera L.

**Moraceae**

 Morus rubra L.

**Nyssaceae**

 Nyssa sylvatica Marshall

**Oleaceae**

Ligustrum sinense Loureiro

**Onagraceae**

Circaea alpina L.

Epilobium coloratum Biehler

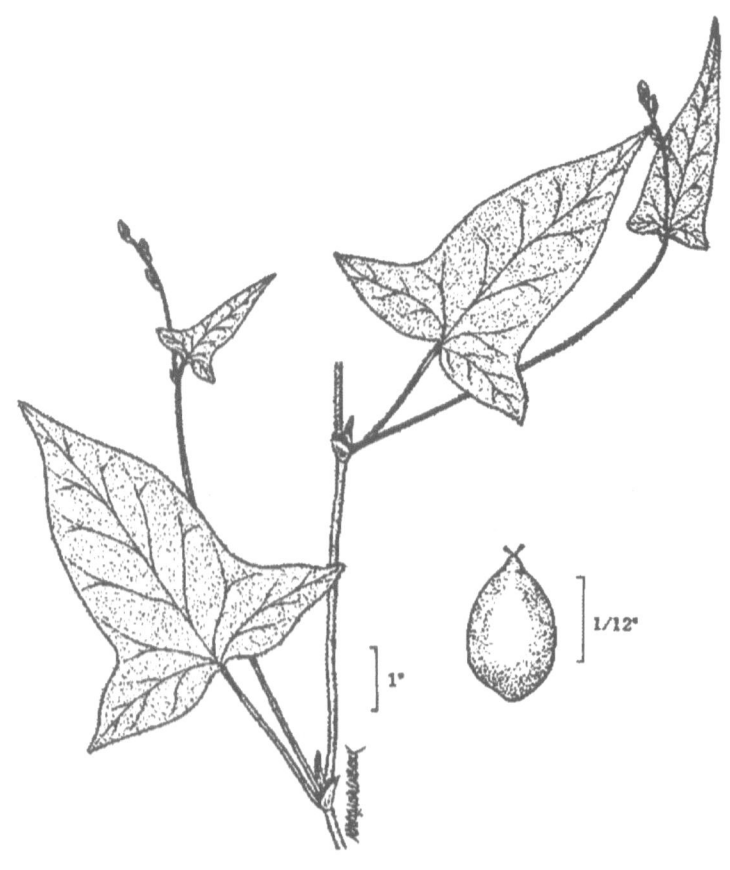

***Polygonum arifolium L.***
Halberdleaf Tearthumb
(illus. fr. USDA NRCS)

Ludwigia alternifolia L.

**Phytolaccaceae**

 Phytolacca americana L.

**Plantaginaceae**

Plantago rugelii Dene.

**Platanaceae**

Platanus occidentalis L.

**Polygalaceae**

 Polygala sanguinea L.

**Polygonaceae**

Polygonum amphibium L.

 P.  arifolium L.

P. caespitosum Bl.

 P. persicaria L.

 P. punctatum Ell.

P. sagittatum L.

P. virginianum L.

***Lysimachia nummularia L.***
Moneywort, Creeping Loosestrife
(illus. fr. Britton & Brown, 1913)

Rumex acetosella L.

## Portulacaceae

Portulaca oleracea L.

## Primulaceae

 Lysimachia ciliata L.

 L. nummularia L.

 L. quadrifolia L.

## Ranunculaceae

Clematis catesbyana Pursh

 C. virginiana L.

Ranunculus abortivus L.

R. acris L.

R. alllegheniensis Britton

R. bulbosus L.

R. fasicularis Muhl.

R. pusillus Poiret

R. recurvatus Poiret

***Thalictrum pubescens Pursh***
King Of The Meadow, Tall Meadow Rue
(clipart courtesy FCIT)

R. repens L.

R. sardous Crantz

R. septentrionalis Poiret

Thalictrum dioicum  L.

 T. pubescens Pursh  (syn. T. polygamum)

Trautvetteria carolinensis (Walter) Vail

## Rosaceae

Crataegus crus-galli L.

Geum canadense Jacquin

G. laciniatum Murray

Physocarpus opulifolius (L.) Maxim

 Rosa multiflora Thunbers

Rosa palustris Marshall

 Rubus hispidus L.

Spiraea alba Du Roi

## Rubiaceae

 Cephalanthus occidentalis L.

 Houstonia caerulea L.

Mitchella repens L.

**Salix sericea Marshall**
Silky Willow
(illus. fr. Britton & Brown, 1913)

Gallium obtusum Bigelow

## Salicaceae

Salix humilis Marshall

 * S. sericea Marshall

## Saxifrigaceae

Saxifriga careyana Gray

S. micranthidifolia (Haw.) Steudel

 S. michauxii Britton

 Tiarella cordifolia L.

## Scrophulariaceae

Chelone glabra L.

 Mimulus ringens L.

Pedicularis canadensis L.

P. lanceolata L.

Penstemon digitalis Nuttall

Veronica americana (Raf.) Bentham

 V. anagallis-aquatica L.

V. peregrina L.

Veronicastrum virginicum (L.) Farwell

***Verbena hastata L.***
Swamp Verbena
(clipart courtesy FCIT)

**Ulmaceae**

Ulmus americana L.

**Urticaceae**

Boehmeria cylindrica (L.) Swartz

Laportea canadensis L.

 Urtica diocia L.

**Verbenaceae**

 Verbena hastata L.

 V. urticifolia L.

**Violaceae**

Viola affinis Le Conte

 V. conspersa Reichenbach

V. cucullata Aiton

V. pubescens Aiton

V. striata Aiton

# SOME ADDITIONAL COUNTY RECORDS

The following county records were also collected but are not considered aquatic.

*Humulus lupulus*

*Amelanchier sanguinea*

*Populus candicans*

*Carex umbellata*

*Bupleurum rotundifolium*

*Onosmodium hispidissimum*

*Triosteum perfoliatum*

*Quercus imbricaria*

*Lillium tigrinum*

*Amphicarpa bracteata*

*Broussonetia papyrifera*

*Orobanche minor*

*Phlox glaberrima*

*Agrimonia pubescens*

## ACKNOWLEDGEMENTS

I would like to thank Clarence Bandy, Beatrice Crouse, John McClintock, Arthur "Smiley" Ratliff and James Reynolds who owned the collecting sites. Bill Arnett and Ray Crouse for helping with the collections, Dr. A. M. Harvill for confirmation and identification of specimens, and D.W. Ogle for supervision of my project. Please excuse any mistakes in this paper; they are my own.

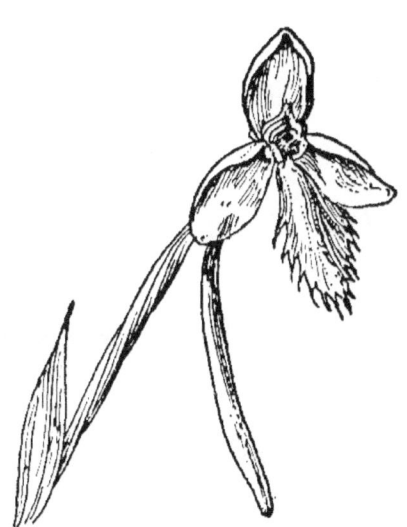

## Inflorescence
*Habenaria lacera*, Ragged Fringed Orchid

## LITERATURE CITED and BIBLIOGRAPHY

BEAL, ERNEST O. 1977. *A Manual of the Marsh and Aquatic Vascular Plants of North Carolina with Habitat Data*. North Carolina State University. Raliegh, N. ,C.

BICKLEY, GEO. W.L.,M.D. 1852. *The History of the Settlement and Indian Wars of Tazewell County, Virginia*. Morgan & Co. Cincinnati, Ohio.

GODFREY, ROBERT K. and Jean W. Wooten. 1981. *Aquatic and Wetland Plants of the Southeastern United States*. University of Georgia Press. Athens, Ga.

BRITTON, N.L. and Addison Brown (1913). *An Illustrated Flora of the Northern United States, Canada and the British Possessions*. Charles Scribner's Sons, Nwe York, NY.

CROUSE, RANDY F. *A Preliminary Floristic Survey of the Aquatic Habitats in Western Tazewell County, Virginia*. Presented at the Spring Collegiate Meeting and Abstracted in Journal of the Tennessee Academy of Science, Volume 59, Number 4, October 1984.

FCIT; FLORIDA CENTER FOR INSTRUCTIONAL TECHNOLOGY, College of Education, University of South Florida, internet http://etc.usf.edu/clipart, accessed 2012.

HARVILL, A. M., JR., Charles E. Stevens & Donna M. E. Ware. 1977, 1985. *Atlas of the Virginia Flora.* Virginia Botanical Associates. Farmville, Va.

MACKENZIE, Kenneth K., 1940. *North American Cariceae.* N.Y. Botanical Garden, pl. 184.

PORTER, H. C., et al. 1948. *Soil Survey of Tazewell County, Virginia.* U.S. Government Printing Office. Washington, D. C.

PRESCOTT, G. W., 1980. *How to Know the Aquatic Plants.* Wm. C. Brown Co.; Dubuque, Iowa.

Soil Survey Staff, Natural Resources Conservation Service, United States Department of Agriculture. *Web Soil Survey.* Available online at http://websoilsurvey.nrcs.usda.gov/. Accessed July 2012.

USDA NRCS. Wetland flora: Field office illustrated guide to plant species. USDA Natural Resources Conservation Service. Provided by NRCS National Wetland Team, Fort Worth, TX.

WAGNER, DAVID F., Soil Survey of Tazewell County, Virginia; United States Department of Agriculture, Natural Resources Conservation Service and the Virginia Polytechnic Institute and State University (Fieldwork by David F. Wagner, Eddie L. Childers, Michael K. Schramm, Thomas D. Adkins, Timothy Anders, and Robert R. Dobos); 2009.

http://soildatamart.nrcs.usda.gov/manuscripts/VA185/0/Tazewell_
VA.pdf. Accessed July 2012.

*Hydrologic Units : Tazewell County*, [Virginia] (185), prepared by
Virginia Department of Conservation and Recreation, Division of
Soil and Water Conservation ; prepared for USDA Natural
Resources Conservation Service, Virginia Department of
Conservation and Recreation, Virginia Department of
Environmental Quality. Ft. Worth, TX : USDA-NRCS-National
Cartographic and GIS Center ; 1995] Rev. January 1995.

***Viola conspersa Reichenbach***
American Dog-violet, Alpine violet
(syn. Viola labradorica Schrank)
(illus. courtesy FCIT)

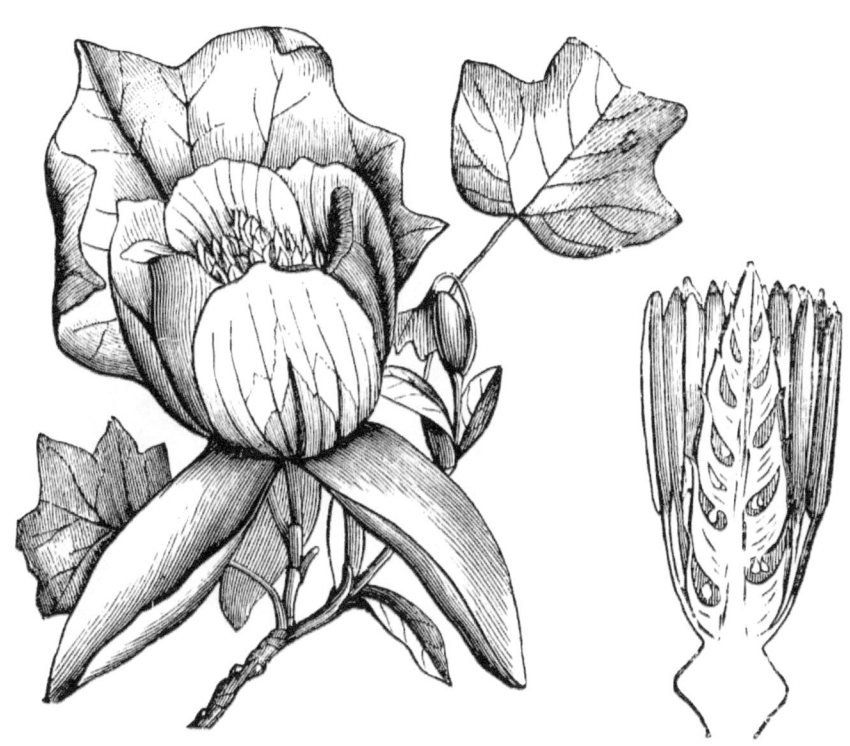

***Liriodendron tulipifera L.***
Tulip Poplar
(illus. curtesy FCIT)

1"

1/12"

2/3

3/5

1/3